40p

Jarrold Fungi Series

Text and photographs by E. A. Ellis

British Fungi **Book 2**
Jarrold Colour Publications, Norwich

This book deals with puffballs and their allies (Gasteromycetes), rusts (Uredinales), smuts (Ustilaginales), cup- and flask-fungi (Ascomycetes), a few moulds and one slime-fungus. The first three groups, like all the species considered in Book 1, are Basidiomycetes. The Gasteromycetes have spores maturing inside the fruit body and in puffballs and earth-stars these spores are intermingled with fine threads (elaters), the dry spores being expelled through a nipple-like aperture or freed through the rupture and peeling of the skin. A few puffballs grow underground, like truffles; some sprout in groups from rotten wood; but most flourish on grassy or sandy ground. Rarities like the little slender-stemmed puffball *Tulostoma brumale* can be found among mosses on calcareous dunes or the tops of old walls and the white *Bovista paludosa* must be looked for in Sphagnum bogs in late summer. Twenty kinds of earth-star have been recorded from Britain; most of them occur in woods or under hedges, but some are associated with calcareous dunes; one very rare species, *Myriostoma coliforme*, has a spore chamber pierced with numerous holes, like a colander. The leathery skinned earth-balls tend to crack and open out on ripening; rain-water collects in them and the spores are dispersed partly by the splashing of raindrops. The fruit bodies of many of these powdery fungi are commonly attacked by certain small beetles. Closely related to the more familiar 'bird's-nests' are several very small species which must be sought in damp, rushy or grassy litter. One of these, *Sphaerobolus stellatus*, is orange-coloured, with a starry cup holding a white-skinned spherical ball of black spores which is ejected like a shot from a gun. In stinkhorns the spore-mass is slimy and glutinous at maturity and becomes exposed on a hood-like receptacle at the tip of a tall stalk which sprouts from a cup-like base (volva). These fungi are noted for disseminating powerful odours to attract insects which are helpful in distributing the spores. One uncommon member of the stinkhorn tribe, *Clathrus ruber*, develops from a jelly-filled 'egg' like the others, but when mature has the appearance of a coral-red lattice-work ball, the meshes of which are coated with slimy, olive-brown spores. This very evil-smelling phalloid grows in hedge bottoms chiefly near the south coast of England and in the Channel Islands. One rare and remarkable Gasteromycete deserving special mention is *Battarraea phalloides*. This emerges from a volva, rather like a stinkhorn, with a rough, tapering, rusty stem up to 30 cm. tall and a cap-like receptacle coated thickly on the outside with reddish-brown powdery spores. It is found sprouting from black, crumbly mould in hollow trees and on old hedge banks where the soils is a mixture of sand and fine leaf-mould, in early autumn. See *A Chronological Catalogue of the Literature of the British Gasteromycetes* by J. T. Palmer (published by J. Cramer, 1968).

All rust-fungi (Uredinales) are parasites. Most of them attack the flowering plants, but hosts also include conifers and ferns. Many go through an elaborate cycle of development in the course of the year, involving the production of a series of different spore-bearing structures and in some cases there is alternating infection of two kinds of host. In the black rust of wheat (*Puccinia graminis*) basidiospores are liberated from over-wintered teleutospores formed on the summer host (in this case wheat or some other grass). When one of these settles on a barberry leaf, it germinates, invading the leaf tissues with mycelium and presently giving rise to a little pustule (pycnidium). Since each basidiospore contains only half of the genetic elements which must combine to produce the mature fungus, the pycnidium developing from it is similarly limited when it buds off masses of little uninucleate pycnidiospores, along with nectar to attact spore-dispersing insects. For further progress to be made, cross inoculation must occur, followed by fusion between the two types of nuclear material. Once this has been achieved, clustercups (aecidia) are formed, bearing chains of binucleate aecidiospores. These are discharged explosively into the air and deposited on the leaves of grasses. Germination is followed by invasion of the leaf cells and the development of spore-bearing pustules. The summer or 'uredo' spores, also binucleate, are produced in great numbers and, being powdery, are widely scattered by the wind. In preparation for winter, thick-walled, dark-coloured teleutospores are formed. The cells are again binucleate, but the contents become fused before basidiospores are produced on their germination in the following spring, with a separation of the sexual elements. In some rusts there are fewer spore forms, see *British Rust Fungi* by M. Wilson and D. M. Henderson (Cambridge University Press, 1966).

The smuts (Ustilaginales) have not imitated the rusts infecting alternative hosts in the course of their development and the number of species is much fewer. In parasitising green plants their presence is often undetected visually until the spores are formed in dark, powdery masses spilling from lesions on leaves and stalks or filling seeds and anthers when the host reaches the reproductive stage. Smuts rely chiefly on their dry, often spiny, chlamydospores (resting-spores) for dissemination by wind. They often are able to survive for several years while awaiting suitable conditions for germination, in a moist atmosphere. When mature, they contain a single diploid nucleus, but on germination a division of the genetic elements takes place whereby haploid promycelial spores are liberated. Recombination of these elements follows before the offspring produce the next crop of chlamydospores. One species worth looking out for is *Thecaphora seminis-convolvuli*, which attacks all four British species of bindweed. Infected flowers are usually smaller than healthy ones and their stamens are much shortened, with the anthers discoloured and bearing bud-spores (conidia) on their surfaces. Later, the chlamydospores of the smut develop, filling the seeds inside the capsules with dark chocolate-brown powder. There are several stripe-smuts conspicuous on the leaves of grasses, while a recent invader, *Ustilago maydis*, produces huge, disfiguring pustules on the heads of maize, now grown as a fodder crop increasingly in this country. Infection in cereal smuts is largely seed-borne. See *British Smut Fungi* by Ainsworth and Sampson (Commonwealth Mycological Institute, 1950).

In Ascomycetes the spores are produced in thin-walled envelopes called 'asci' and, when ripe, extruded through a pore at the tip or by rupture of the envelope. Most commonly, though by no means always, asci contain eight spores. These fungi very often have supplementary means of reproduction involving the development of organs bearing asexual spores (conidia). The connection between some conidial forms and their ascus-bearing counterparts in many cases awaits recognition and such forms are known as 'fungi imperfecti', being named and classified either as Coelomycetes (with well-defined fruit bodies at first enclosing the spores) or Hyphomycetes (with spores produced openly on thready or tufted conidiophores). Many small leaf-spotting fungi and moulds are included in these two categories. To achieve ultimate perfection, the fusion of sexual elements must be achieved at some stage. The Ascomycetes comprise very numerous species with fleshy, cup-shaped fruit bodies (apothecia) and others with separate or thickly massed flask-shaped bodies (perithecia). Among the former are some relatively large species found growing from soil or leaf-mould and having asci whose spores pop out through a lidded aperture (these are known as 'operculate discomycetes'). In others (inoperculate) there is no lid or plug at the mouth of the ascus and most species grow from dead wood, stems, leaves and fallen fruits, some being parasitic. Even non-parasites tend to be associated with particular plants and they play an important part in reducing defunct vegetation to friable humus. The flask-fungi (Pyrenomycetes) also have inoperculate asci, which are enclosed in spherical or flask-shaped chambers from which the massed spores are eventually extruded through a common aperture. Often their fruits, which may be of a hard, horny texture or soft and translucent, are grouped in crusts on rotten wood, stems of herbaceous plants or the decaying remains of woody fungi. In many cases they form spots on living or dead leaves and stems, some being parasitic. They include several species which attack insects and spiders and mature after they have killed their victims. There is almost no limit to the enterprise shown by these fungi in finding niches in which to flourish. See *British Ascomycetes* by R. W. G. Dennis (published by J. Cramer, 1968). Lichens are omitted from consideration here, but it should be mentioned in passing that there is an ascomycete component linked commensally with an alga in each of these organisms. As stated earlier, many Ascomycetes at some stage produce asexual spores (conidia). Sometimes conidial fungi are found unconnected with known ascophorous species. To cope with this situation a separate system has been devised for classifying and naming them. See *British Stem and Leaf Fungi* (*Coelomycetes*) by W. B. Grove (Cambridge University Press, 1935–7) and *Dematiaceous Hyphomycetes* by M. B. Ellis (Commonwealth Mycological Institute, 1971). Mention must also be made of the remarkable assortment of water-moulds which develop on submerged, decaying leaves in our streams and lakes. These fascinating little fungi have been fully described and illustrated in *Aquatic and Water-borne Hyphomycetes* by C. T. Ingold (Freshwater Biological Association, 1975). An indispensable work of general reference for both budding and professional mycologists must always be Ainsworth and Bisby's *Dictionary of the Fungi*, published in a succession of editions by the Commonwealth Mycological Institute.

It is hoped that by presenting some of the attractions of British fungi on these pages a wider interest in them will be encouraged in those who explore and wish to enjoy in every possible way the natural wonders of our country-side.

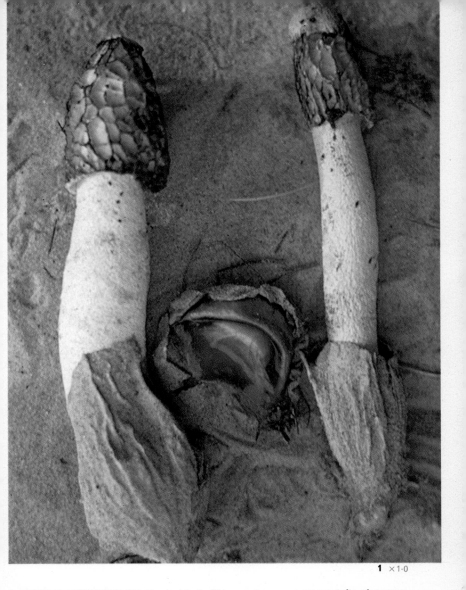

1 ×1·0

1. DUNE STINKHORN (*Phallus hadriani*). This species grows among tufts of marram grass in the younger dunes along some parts of our coast, chiefly in East Anglia, appearing most frequently in autumn. It differs from the common woodland stinkhorn in the violet tinting of its 'eggs' and in its hyacinth-like scent.

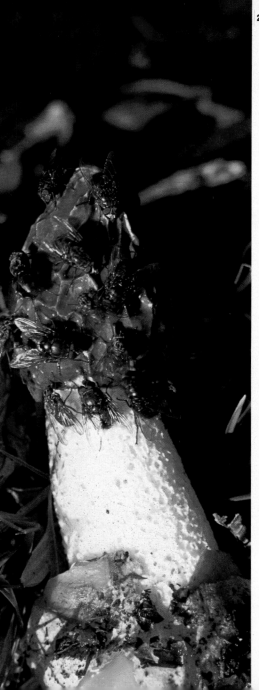

2. COMMON STINKHORN or WOOD-WITCH (*Phallus impudicus*)

with flies on cap. It is a common experience when taking a walk through woodlands in mild, showery weather, to notice a half-sweet, half-fetid odour rising from the undergrowth. This is the carrion scent emitted by stinkhorns to attract flies to the olive-brown, slimy coating of spores on their conical caps. Although the insects consume most of this feast very quickly, leaving a white, honeycombed cone exposed, they help to disperse spores which stick to their legs and get carried away. In many cases spores pass undigested through their alimentary systems, as pips from berries pass unharmed through birds. These fungi grow on very rotten wood and leaf-mould impregnated with their white, thready 'spawn'. Clusters of white-skinned young fruit bodies resembling turtles' eggs are formed and these have a sort of 'yolk' surrounded by greenish-brown jelly. When rain soaks the ground the skin is ruptured and a long, cylindrical white stem bearing its cone of dark slime springs up from each egg, expanding very swiftly. The light, spongy, rather fragile columns have an airy, cellular structure and look rather like gas-mantles, especially when they have lost their spores. The smaller **DOG STINKHORN** (*Mutinus caninus*) has a yellow stem and an orange-red cap coated with dark green, sticky spores. It is most often found in birch woods, but sometimes appears on strawy manure-heaps. Fungi of this type are common in tropical forests where some of them are adorned with elaborate lacy frills and collars. A few alien species have turned up from time to time in English hot-houses. All emit carrion-like odours.

3. BIRD'S-NEST FUNGUS (*Cyathus striatus*).

These are so called from their resemblance to miniature nests containing clutches of little egg-like bodies.

This species forms clusters on rotten twigs and pine needles on the floors of gloomy woodlands. The cups are dark brown and shaggy outside and bluish white within. The mouths are closed at first with creamy-white membranes which become gelatinous and dissolve to reveal the whitish 'eggs'. These are packed with spores like those of puff-balls. Each 'egg' sits on a coiled gelatinous thread and when the cup fills with water in rainy weather it rises to the top and is apt to be splashed out by the impact of raindrops. Some may stick to the feet of passing animals, such as rabbits or foxes, or may be carried far afield by birds which come in contact with them; even crawling insects and molluscs may act as dispersal agents.

4. FIELD BIRD'S-NEST (*Cyathus olla*).

This is the commonest member of its tribe, appearing on the soil of arable fields and gardens littered with dead leaves or straw in late summer and autumn. It is broadly trumpet-shaped and smooth inside.

3 ×1·0
4 ×1·5

5 ×1·0
6 ×0·5

5. ROSY EARTH-STAR (*Geastrum rufescens*). Earth-stars are closely related to puffballs. They differ in having a thick outer rind which eventually splits and spreads outwards to form pointed lobes in the form of a star, with the puffball-like portion exposed in the centre. They appear in autumn, often in groups, under trees and bushes along hedges and in woods, especially where the soil is sandy. About seventeen kinds have been noticed in Britain. *G. rufescens* occurs locally in woods and is distinguished by the pink colour when fresh.

6. COMMON EARTH-STAR (*Geastrum triplex*). This species may often be found plentifully in woods where there is deep leaf-litter. Young specimens look like tulip bulbs and on expansion the peel tends to split in such a way as to form a thick 'collar' between the points and the puffer. Spores issue like smoke from the small mouth at the top when their container is hit by raindrops.

7. FOUR-RAYED EARTH-STAR (*Geastrum quadrifidum*). A rather rare pine-wood fungus with four stiff, leathery 'legs', a stalked spore case (peridium) and a cup-like base to which the points of the star remain attached where it is embedded in the leaf-litter. Some earth-stars, especially certain kinds growing on dunes and calcareous grassland, dry out and are scattered and rolled along by the winds in winter, after the fashion of tumble-weeds, scattering spores as they go. In the very small *G. recolligens* (*nanum*), a rare species found on sandy soils, the rays (silvery beneath) curl upwards when dry to cover the spore-body completely, making wind transport effective.

8. DUSKY PUFFBALL (*Bovista plumbea*). This and the larger, almost black *B. nigrescens* are spherical, stemless, smooth-skinned puffballs of pastures, downs and dunes. When ripe, they often become detached and blown about. Well-preserved specimens of the larger kind have been found in prehistoric and Roman rubbish-tips in various parts of Britain; they may have been used medicinally, or as tinder.

9. SPINY PUFFBALL (*Lycoperdon foetidum*). Found on grassy heaths, this has dark spines which are eventually lost, leaving hexagonal markings. It is sometimes regarded as a variety of the **COMMON PUFFBALL** (*L. perlatum*).

10. COMMON EARTH-BALL (*Scleroderma aurantium*). Earth-balls have thicker, coarser skins than puffballs and resemble truffles in their solid flesh when young. This species can be found very commonly on sandy heaths and in birch woods in autumn.

11. GIANT PUFFBALL (*Calvatia gigantea*). Often as large as a man's head, these 'Bulfers' flourish in wayside nettle-beds in July and August. When young and white they are pleasantly edible, but on darkening the flavour deteriorates. The fluffy matrix of dried specimens has long been used as a styptic.

8 ×1·0

9 ×0·75

10 ×0·5

11 ×0·14

12. REED-and-DOCK RUST *(Puccinia phragmitis).* Docks growing in marshes commonly develop crimson stains on their leaves in May and June and the undersides of the discoloured patches bear groups of little crater-like, white clustercups of a rust which in turn infects young leaves of the common reed. Long, narrow cushions (sori) of brown uredospores and darker teleutospores develop on the reeds in summer and autumn. The teleutospores germinate and reinfect docks in spring.

13. MARSH MARIGOLD RUST *(Puccinia calthaecola).* The whole life-cycle of this parasite is spent on the same host species. The illustration shows yellow clustercups (aecidia) appearing in spring and small brown uredo-sori developing on a young leaf. Thick-walled, dark, two-celled teleutospores appear later in the season. *P. calthae* occurring on the same host is distinguished by its slender, pale brown, long-stalked teleutospores.

14. GROUNDSEL CLUSTERCUP RUST *(Puccinia lagenophorae).* The Common Groundsel of Europe was introduced in Australia by early colonists. There it became parasitised by a rust of local, daisy-like *Lagenophora* species. Eventually, in 1961, this fungus appeared on groundsel in Europe and has since become widespread. It also attacks Oxford ragwort and can infect marigolds and common daisies.

15. OXALIS RUST *(Puccinia oxalidis).* This newcomer to Europe and Britain appears to be of South American origin. First noticed on the Mexican *Oxalis latifolia* in the Channel Islands in 1973, it has since appeared on this and the pink bulbous *Oxalis corymbosa* in various parts of England where these plants are troublesome weeds. In America it is said to have aecidia on leaves of certain barberries, but it is able to persist here from year to year on oxalis, especially where the climate is mild through the winter.

12 ×2·0

13 ×1·5

14 ×3·0

15 ×1·5

16. POPLAR-and-DOG'S MERCURY RUST (*Melampsora populnea*). The orange-yellow aecidial stage of this fungus is shown here on leaves of dog's mercury. It occurs commonly on this host in spring. In other strains of the same rust, aecidia develop on needles of European larch and some pines. Later in the year the fungus goes on to attack aspen and white poplar, first producing small, orange-coloured powdery sori of uredospores and then dark teleutospores underneath the leaves.

17. MEADOWSWEET RUST (*Triphragmium ulmariae*). The bright orange, powdery aecidiospores of this common rust appear in conspicuous masses, distorting the stems and veins of meadowsweet leaves in early summer. Smaller, pale yellow groups of uredospores develop on the undersides of the leaves later, followed by brown-black, three-celled, warty, triangular teleutospores which persist through the winter. A similar but rare rust (*T. fili-pendulae*) attacks dropwort in chalk country.

16 ×2·0   **17** ×0·5

18. COMFREY RUST (*Melampsorella symphyti*). Although the aecidia of this rust on needles of silver fir (*Abies alba*) have not yet been detected in Britain, infected stocks of wild and cultivated comfreys bear orange uredospores on the undersides of their leaves year after year, though by no means commonly. The mycelium is perennial in the host. The fungus flourishes particularly on the native comfrey (*Symphytum officinale*) in some of our East Anglian fens.

19 ×2·0

20 ×1·0
21 ×1·5

19. RESIN-TOP RUST *(Peridermium pini)*. The mycelium of this fungus persists in diseased branches of Scots pine, producing cankers and large, orange, bladder-like cluster-cups annually in May and June. The trees tend to die after a time. This self-perpetuating race of the rust appears to be indigenous in Scotland and has spread to pine plantations in England and Ireland in recent years. Other races *(Cronartium flaccidum)* have a non-peren-nating aecidial stage on pines and later stages on a wide range of herbaceous plants.

20. JUNIPER-and-HAWTHORN RUST *(Gymnosporangium clavariiforme)*. During rainy weather in spring, swollen, cankered boughs of wild and cultivated junipers can be seen clothed with conspicuous orange spikes. These are soft, gelatinous masses of teleuto-spores produced by the perennial mycelium of this rust. The fungus is also capable of infecting leaves and fruits of hawthorn on which yellow clustercups develop during the summer.

21. BRAMBLE RUST *(Phragmidium violaceum)*. This rust flourishes almost everywhere on many of the commoner sorts of blackberry, being most noticeable in autumn, when it causes crimson and purple spotting on the leaves. The powdery aecidia appear as orange-yellow lesions on stems and leaf veins in May and smaller patches of orange uredospores appear on the underside of the leaves, followed by black clusters of teleutospores (usually four-celled).

22. PURPLE ANTHER SMUT *(Ustilago violacea)*. The anthers of campions and many of their relatives are commonly parasitised by this smut-fungus, whose purplish-brown, powdery spores usurp the place of pollen. As in the sea campion illustrated, affected flowers appear dark and dusty in their centres. The fungus mycelium has a permanent hold and every flower on a diseased plant develops smut spores.

23 ×0·75

24 ×0·4
25 ×0·5

23. SLIMY LEOTIA (*Leotia lubrica*). A fairly common autumnal fungus on mossy banks, often with bracken, usually appearing in troops. The fruit bodies are honey-yellow or olive-green and have granular yellow stalks. The heads become very slimy in wet weather. There is one other British species, *L. atrovirens*, which is entirely dark blue-green; this is found only rarely, in mossy woods.

24. HALF-FREE MOREL (*Mitrophora hybrida*). This fleshy ascomycete sprouts from moist, black woodland soil early in May. It is sometimes plentiful under poplars and old hawthorns. It is distinguished from other morels by the free margin of the small, conical cap, which is yellowish brown with crinkled black ribs. It varies much in size and the hollow stem may be cylindrical or bulging either above or below.

25. COMMON MOREL (*Morchella esculenta*). This relative of the cup-fungi has an irregularly honeycombed top and a stout, wrinkled stem. It can be found on bare soil under trees and hedges, often in gardens, in April and May and is commonest in limy districts. It is a popular edible species throughout Europe. The fungus should be washed thoroughly before cooking, to remove gritty soil from its numerous cavities. It is often served with lemon juice.

26. SCARLET CUP (*Sarcoscypha coccinea*). Sometimes called 'fairies' bath', this gorgeous fungus sprouts from mossy twigs rotting on the ground in damp places and is usually at its best in February. It develops most commonly on hazel sticks in hollows of old woodlands, but also appears on fallen branches of sallow and alder where ground is often flooded. The cups are usually stalked, with the upper surface and the underside coated with white or faintly rose-pink down. A pure white form is found very occasionally.

27 ×1·0

28 ×2·9

29 ×2·0

30 ×1·25

31 ×0·75

27. GOLDEN EAR (*Otidea onotica*). Most fleshy cup-fungi growing on the ground are brown and broadly rounded. This species, developing in clusters among fallen oak leaves in autumn, has ear-shaped cups split down one side and is peach-coloured, often with a flush of pink. Some specimens are 10 cm. high.

28. SEDGE-CUP (*Sclerotinia sulcata*). The stalked fruits (apothecia) of this fungus sprout from black, ribbed, spindle-shaped sclerotia embedded in the stalks of sedges in marshes in spring. Their spores infect young flowering stems, which die back and fail to yield seeds. Sclerotia (hard, black resting-bodies) are formed and survive the winter and give rise to stalked cups in April.

29. CATKIN-CUP (*Ciboria amentacea*). These long-stemmed cups, shaped like wine-glasses, can be found commonly under alders and sallows in early spring. They arise from the previous year's fallen male catkins and emit clouds of spores which infect the new crop as it falls to the ground.

30. SAND STAR-CUP (*Peziza ammophila*). When sand-dunes become soaked with rain in autumn, certain fungi spring up where mycelium has developed on buried fragments of vegetation. The curious star-cup has a rooting, stem-like base and expands with starry lobes pushing back the sand, near tufts of marram grass.

31. TAR-SPOT of SYCAMORE (*Rhytisma acerinum*). Round black patches conspicuous on sycamore leaves in summer represent an early stage in the growth of this cup-fungus. At first, only conidia are formed; the grey-crusted apothecia mature after the leaves have fallen and lain on the ground through the winter.

32. ERGOTS of *Claviceps purpurea* **on RYE.** The horn-like sclerotia of this fungus develop on the ears of many wild grasses and cereals. They contain poisons which in the past have caused gangrene and madness in people eating bread made from diseased rye and are a cause of abortion in cattle; but their products have valuable medical uses. Over-wintered sclerotia produce purplish, round-headed clubs bearing ascospores in crowded perithecia.

33. WHITE-STALKED SADDLE (*Leptopodia albella*). Several kinds of long-stalked cup-fungi with saddle-shaped heads grow from woodland soil in summer and autumn. *L. albella*, with its orange-brown top and rather slender white stalk, found in Norfolk in 1970, is a recent addition to the British fungus flora.

34. GREY SADDLE (*Helvella lacunosa*). The stalks in *Helvella* are hollowed and furrowed and the saddle-like caps are often contorted. The flesh in this autumnal species is smoky brown or grey, with a lavender bloom at maturity. It grows on the ground in woods. The commoner *H. crispa* is cream coloured, with a two-lobed cap. It contains helvellic acid, which dissolves red blood corpuscles. This is present also in the large, reddish-brown, morel-like *Gyromitra esculenta* found under pines in early spring; both fungi are safe to eat when cooked.

33 ×0·75
34 ×1·0

35 ×0·5

36 ×0·75   **37** ×0·75

35. ORANGE-PEEL CUP *(Aleuria aurantia)*. This is a common species, appearing in plenty along woodland tracks, road verges and garden corners throughout the autumn. Its large, often unevenly shaped cups, orange-red inside and whitish and downy outside, are rather thin and brittle, often assuming a flattened form when fully expanded. On being touched they are apt to eject spores like puffs of smoke.

36. WOODLAND GROUND-CUPS *(Peziza michelii* and *P. succosa)*. Both of these grow on bare soil in woods, chiefly in late summer and autumn. The former is reddish brown, with very neat, perfectly round cups when young and is a rather uncommon species of damp woods. *P. succosa*, similar in form, is much lighter in tint and greyish or yellowish; it is common on clayey soil under trees.

37. GREEN WOOD-CUP *(Chlorosplenium aeruginascens)*. This is the commonest cause of the vivid blue-green staining often noticeable in dead oak wood which has lost its bark. The mycelium forms the green pigment (xylindene) which also suffuses the spore-bearing cups when they mature on wet, fallen boughs. The stained wood has often been employed in marquetry and veneering and is a feature of the famous Tunbridge ware. When polished, it has the beauty of jade. Ash wood is often stained by the same fungus. Another species, *C. aeruginellum*, gives a blue-green tint to withered stems of meadowsweet.

38. NUT-CUP *(Hymenoscyphus fructigenus)*. Many acorns and hazel nuts falling to the ground in woods fail to grow because they are damaged by insects, and they lie rotting gradually in the leaf-litter. In the following autumn they commonly bear crops of this little ivory-tinted cup-fungus, especially in damp, dark places. This species also appears, with other fungi, on old beech cupules and on the small nuts of hornbeam.

38 ×1·25

39 ×3·0
40 ×1·0

39. HOGWEED SEED-CUP (*Symphyosirinia* species). This parasitic fungus, discovered in Scotland in 1969, springs in late summer and autumn from the previous year's fallen mericarps of hogweed on damp soil. Newly fallen fruits first bear pale pink, sessile heads of many-septate, cylindrical, slimy conidia. Long-stalked clubs bearing similar spores develop in the following year, together with mature apothecia. Other species with the same type of development attack fruits of angelica, milk parsley and marsh bedstraw. This last species, *S. galii*, has conidia topped with three long, elegant spines.

40. ALDER LEAF-CURL (*Taphrina tosquinetii*). A common parasite of alder, crumpling and discolouring the leaves in summer. The fungus forms asci in compact, thin crusts on the leaf surface. Various species of *Taphrina* produce 'witches' brooms' on trees and one (*T. deformans*) is responsible for the destructive red leaf-curl disease of peaches and almonds.

41. SCARLET CATERPILLAR-CLUB (*Cordyceps militaris*). This is the most conspicuous of several fungi which are parasitic on insects in this country. Its fertile clubs can be found sprouting from soil and mossy turf at almost any time of the year, but careful probing reveals that in every case they are growing from buried caterpillars or pupae of moths.

42. 'CHOKE' of GRASSES (*Epichloe typhina*). Commonly sheathing the stems of grasses, this appears first as a thin white crust, later thickening as the orange perithecia develop. Often conspicuous on cock's-foot grass, it also attacks many other species. Diseased shoots usually fail to produce flowers. The fungus is often devoured by larvae of a fly, *Chortophila spreta*, an insect like a miniature house-fly.

41 ×3·0
42 ×1·5

43 ×10·0
44 ×0·75

43. CUP NECTRIA (*Nectria peziza*). The translucent yellow perithecia of this species develop in swarms on soft, decayed bracket-fungi such as the Scaly Polypore and on wet, rotten wood. When fresh, they are spherical, with spores escaping through a small mouth at the top, but later, they collapse and appear like shallow, brownish cups.

44. CRAMP-BALL (*Daldinia concentrica*). The trunks of dead ash trees are often studded with brown or shiny black knobs. These are fruit bodies of this fungus. In section they look like charcoal, with concentric rings. The interior contains jelly-like material. Black spores are produced in thousands of funnel-like chambers near the surface. Pressure built up in the jelly is used for ejecting the spores at night. This happens throughout the summer, after which the fruit-body is exhausted.

45. FALSE TRUFFLE (*Elaphomyces muricatus*). This thick-skinned, roughly warted, tuber-like subterranean fungus is found chiefly in pine woods, a few inches beneath the surface. It is brownish purple in section, with a marbling of white veins. There is a dark powdery 'core'. The edible truffle most frequently found in Britain is darker brown, with a coarsely, hexagonally cracked surface and occurs mainly in chalky soil under beeches. False truffles are parasitised by two species of *Cordyceps* (*C. capitata* and *C. ophioglossoides*).

46. WITCHES' FINGERS (left to right: *Xylosphaera longipes* and *X. polymorpha*). These firm black or dark brown fungi sprout from dead tree-stumps and fallen branches. *X. longipes* occurs chiefly on sycamore and has long slender clubs, at first powdered with white. *X. polymorpha* grows at soil-level from trunks of various trees and has plump, finger-like clubs with short stems.

45 ×2·0

46 ×1·0

47. PIN-MOULD on an AGARIC. These moulds (Mucoraceae) have a delicate, branched mycelium feeding and growing rapidly on dung, rotten fruit and the like. They also give rise to globular vessels (sporangia) containing black spores, at the tips of tall, colourless stalks, resembling so many pins with at first yellow and finally black heads.

48. SOOTY MOULD (*Arthrinium sporophleum*) on a faded sedge leaf. This is one of the innumerable dark-spored hyphomycetes (Dematiaceae) found on living and dead plants and their products. It forms sooty spots on newly dead sedges and rushes. The colonies are composed of transparent threads with dark septa, bearing along their whole length numerous olive-black, bluntly spindle-shaped spores.

49. FLY-MOULD (*Entomophthora* species) on **HOVER-FLIES** (*Syrphus balteatus*). Species of *Entomophthora* cause epidemic disease in a variety of insects, including flies, beetles, sawflies, aphids, earwigs and caterpillars of moths and butterflies, while a few attack mites and harvest-spiders. The house fly is a common victim, as are the small hover-flies which often swarm in the countryside in summer. These fungi destroy countless aphids. See *Key to the species of Entomophthora*, by G. M. Waterhouse, *Bulletin of the British Mycological Society* 9,1 (1975).

48 ×2·0
49 ×2·0

50. BARK 'PUFFBALL' (*Reticularia lycoperdon*). Soft, milk-white excrescences often appear on dead trunks of willow, alder and birch in early spring. These soon develop a shining silvery skin and, when ripe, gradually peel and disintegrate as a mass of dark brown, powdery spores (as in puffballs). They are the fruit bodies of one of the larger slime-fungi (Myxomycetes).